**NASA
Reference
Publication
1383**

April 1996

Total Solar Eclipse
of 1998 February 26

Fred Espenak
and Jay Anderson

ECLIPSE PREDICTIONS

INTRODUCTION

On Thursday, 1998 February 26, a total eclipse of the Sun will be visible from within a narrow corridor which traverses the Western Hemisphere. The path of the Moon's umbral shadow begins in the Pacific, continues through northern South America and the Caribbean Sea (Figure 3), and ends at sunset off the Atlantic coast of Africa. A partial eclipse will be seen within the much broader path of the Moon's penumbral shadow, which includes parts of the United States and eastern Canada, Mexico, Central America and the northern half of South America (Figures 1 and 2).

UMBRAL PATH AND VISIBILITY

The Moon's umbral shadow first touches down on Earth just south of the equator in the Pacific Ocean about 3000 kilometers southeast of the Hawaiian Islands (15:46:45 UT). Even at this extreme position, the total eclipse lasts nearly one and a half minutes. For the first one and a quarter hours, the umbra sweeps across 5500 kilometers of open water while encountering no major bodies of land. First landfall finally occurs shortly after crossing north of the equator (16:56 UT) as the shadow rushes across several members of the Galápagos Islands. Maximum eclipse takes place shortly before noon with the Sun 69° above the horizon. Unfortunately, none of the islands are on the center line where the duration of totality is one second under four minutes (Figure 4). The northern third of Isla Isabela lies within the path and experiences a duration of 3 minutes 21 seconds along its north coast. Isla Pinta and Isla Marchena are also situated in the path, but all three islands are uninhabited wildlife sanctuaries with access strictly controlled by the Ecuadorian government.

After leaving the Galápagos, the umbral path continues on a northeastern course. The instant of greatest eclipse[1] occurs at 17:28:23.2 UT about 600 kilometers west of Colombia's Pacific coast. At that moment, the length of totality reaches its maximum duration of 4 minutes 8 seconds, the Sun's altitude is 76°, the path width is 151 kilometers and the umbra's velocity is 0.592 km/s. The axis of the Moon's shadow passes closest to the center of Earth (gamma[2] =0.239) at greatest eclipse.

Fifteen minutes later (17:43 UT), the umbral shadow enters South America and follows the border between Panama and Colombia (Figure 5). Stretching across the Isthmus of Panama, the path crosses into northern Colombia where the center line duration drops below four minutes. In spite of a 25 second penalty for its position 20 kilometers north of the center line, the Colombian city of Valledupar still experiences a generous 3 minute 29 second total eclipse. After climbing the Andes, the path sweeps through northwestern Venezuela (Figure 6). The semi-desert vegetation of this region is a testament to its arid climate. Although Maracaibo lies 50 kilometers south of the center line, it manages to enjoy nearly three minutes of totality. The eclipse occurs here in the early afternoon with the Sun 65° above the horizon.–

The Moon's shadow leaves Venezuela via the Peninsula de Paraguaná and sweeps across the Caribbean where it encounters five major islands of the Lesser Antilles. The center line passes directly between Aruba and Curaçao (Figures 7, 8 and 9), but both islands experience as much as 3 1/2 out of a possible 3 3/4 minutes of totality. Aruba's capital Oranjestad witnesses a 3 minute 6 second total phase while the duration in Willemstad, Curaçao is 1 minute 58 seconds. Both Netherlands Antilles islands make good locations for the eclipse since they share similarly promising weather prospects. Nearby Bonaire is just outside the 143 kilometer wide path although its northwestern coast will witness a grazing eclipse with perhaps a dozen seconds of totality.

The track continues across the Caribbean where it engulfs the Leeward Islands of Montserrat, Antigua and Guadeloupe (Figures 10 through 13). Here, the umbra's velocity increases to over 0.9 km/s and the path width drops to 136 kilometers. Occurring in mid afternoon, the Sun's altitude is 49° at maximum eclipse. Although the center line passes between Guadeloupe and the other two islands, all locales enjoy as much as 3 out of a possible 3 1/4 minutes of totality. Since the southern extremes of Guadeloupe actually

---

[1] The instant of greatest eclipse occurs when the distance between the Moon's shadow axis and Earth's geocenter reaches a minimum. Although greatest eclipse differs slightly from the instants of greatest magnitude and greatest duration (for total eclipses), the differences are usually negligible.

[2] Minimum distance of the Moon's shadow axis from Earth's center in units of equatorial Earth radii.

lie outside the umbral path, most observers will want to position themselves on the north and west coasts in order to experience the maximum duration possible. Nevertheless, some may choose a site several kilometers inside the southern limit in order to witness the prolonged views of Baily's Beads and chromosphere afforded by such locations. While not as promising as Aruba and Curaçao, the weather prospects of this region are still quite favorable.

Leaving the Caribbean Sea, the umbra races across the Atlantic towards Africa. About 1000 kilometers west of Morocco, the path ends along the sunset terminator as the shadow falls back into space (19:09:57 UT). Over course of 3 hours and 23 minutes, the Moon's umbra travels along an approximately 14000 kilometer long path and covers a region comprising 0.3% of Earth's surface.

GENERAL MAPS OF THE ECLIPSE PATH

ORTHOGRAPHIC PROJECTION MAP OF THE ECLIPSE PATH

Figure 1 is an orthographic projection map of Earth [adapted from Espenak, 1987] showing the path of penumbral (partial) and umbral (total) eclipse. The daylight terminator is plotted for the instant of greatest eclipse with north at the top. The sub-Earth point is centered over the point of greatest eclipse and is indicated with an asterisk-like symbol. The sub-solar point at that instant is also show.

The limits of the Moon's penumbral shadow define the region of visibility of the partial eclipse. This saddle shaped region often covers more than half of Earth's daylight hemisphere and consists of several distinct zones or limits. At the northern and/or southern boundaries lie the limits of the penumbra's path. Partial eclipses have only one of these limits, as do central eclipses when the shadow axis falls no closer than about 0.45 radii from Earth's center. Great loops at the western and eastern extremes of the penumbra's path identify the areas where the eclipse begins/ends at sunrise and sunset, respectively. If the penumbra has both a northern and southern limit, the rising and setting curves form two separate, closed loops. Otherwise, the curves are connected in a distorted figure eight. Bisecting the 'eclipse begins/ends at sunrise and sunset' loops is the curve of maximum eclipse at sunrise (western loop) and sunset (eastern loop). The exterior tangency points **P1** and **P4** mark the coordinates where the penumbral shadow first contacts (partial eclipse begins) and last contacts (partial eclipse ends) Earth's surface. If the penumbral path has both a northern and southern limit (as does the 1998 February eclipse), then the interior tangency points **P2** and **P3** are also plotted and correspond to the coordinates where the penumbral cone becomes internally tangent to Earth's disk. Likewise, the points **U1** and **U2** mark the exterior and interior coordinates where the umbral shadow initially contacts Earth (path of total eclipse begins). The points **U3** and **U4** mark the interior and exterior points of final umbral contact with Earth's surface (path of total eclipse ends).

A curve of maximum eclipse is the locus of all points where the eclipse is at maximum at a given time. They are plotted at each half hour Universal Time (UT), and generally run from northern to southern penumbral limits, or from the maximum eclipse at sunrise or sunset curves to one of the limits. The outline of the umbral shadow is plotted every ten minutes in UT. Curves of constant eclipse magnitude[3] delineate the locus of all points where the magnitude at maximum eclipse is constant. These curves run exclusively between the curves of maximum eclipse at sunrise and sunset. Furthermore, they are parallel to the northern/southern penumbral limits and the umbral paths of central eclipses. Northern and southern limits of the penumbra may be thought of as curves of constant magnitude of 0%, while adjacent curves are for magnitudes of 20%, 40%, 60% and 80%. The northern and southern limits of the path of total eclipse are curves of constant magnitude of 100%.

At the top of Figure 1, the Universal Time of geocentric conjunction between the Moon and Sun is given followed by the instant of greatest eclipse. The eclipse magnitude is given for greatest eclipse. For central eclipses (both total and annular), it is equivalent to the geocentric ratio of diameters of the Moon and Sun. Gamma is the minimum distance of the Moon's shadow axis from Earth's center in units of equatorial Earth radii. The shadow axis passes south of Earth's geocenter for negative values of Gamma. Finally, the Saros series number of the eclipse is given along with its relative sequence in the series.

---

[3] Eclipse magnitude is defined as the fraction of the Sun's diameter occulted by the Moon. It is strictly a ratio of *diameters* and should not be confused with eclipse obscuration which is a measure of the Sun's surface *area* occulted by the Moon. Eclipse magnitude may be expressed as either a percentage or a decimal fraction (e.g.: 50% or 0.50).

STEREOGRAPHIC PROJECTION MAP OF THE ECLIPSE PATH

The stereographic projection of Earth in Figure 2 depicts the path of penumbral and umbral eclipse in greater detail. The map is oriented north up with the point of greatest eclipse near the center. International political borders are shown and circles of latitude and longitude are plotted at 20° increments. The region of penumbral or partial eclipse is identified by its northern and southern limits, curves of eclipse begins or ends at sunrise and sunset, and curves of maximum eclipse at sunrise and sunset. Curves of constant eclipse magnitude are plotted for 20%, 40%, 60% and 80%, as are the limits of the path of total eclipse. Also included are curves of greatest eclipse at every half hour Universal Time.

Figures 1 and 2 may be used to quickly determine the approximate time and magnitude of maximum eclipse at any location within the eclipse path.

EQUIDISTANT CONIC PROJECTION MAP OF THE ECLIPSE PATH

Figure 3 is an equidistant conic projection map chosen to minimize distortion, and which isolates a specific region of the umbral path. Once again, curves of maximum eclipse and constant eclipse magnitude are plotted and labeled. A linear scale is included for estimating approximate distances (kilometers). Within the northern and southern limits of the path of totality, the outline of the umbral shadow is plotted at ten minute intervals. The duration of totality (minutes and seconds) and the Sun's altitude correspond to the local circumstances on the center line at each shadow position.

The scale used in Figure 3 is approximately ~1:10,000,000. The positions of larger cities and metropolitan areas in and near the umbral path are depicted as black dots. The size of each city is logarithmically proportional to its population using 1990 census data (Rand McNally, 1991).

EQUIDISTANT CYLINDRICAL PROJECTION MAPS OF THE ECLIPSE PATH

Figures 4 through 13 all use a simple equidistant cylindrical projection scaled for the central latitude of each map. They all use high resolution coastline data from the World Data Base II (WDB) and World Vector Shoreline (WVS) data bases and have scales of 1:2,500,000 or higher. These maps were chosen to isolate small regions of the eclipse path of particularly high interest. Once again, curves of maximum eclipse and constant eclipse magnitude are included as well as the outline of the umbral shadow. A special feature of these maps are the curves of constant umbral eclipse duration (i.e.: totality) which are plotted within the path. These curves permit fast determination of approximate durations without consulting any tables. Furthermore city data from a newly enlarged geographic data base of over 90,000 positions are plotted to give as many locations as possible in the path of totality. Local circumstances have been calculated for all positions shown and can be found in Tables 9-22.

ELEMENTS, SHADOW CONTACTS AND ECLIPSE PATH TABLES

The geocentric ephemeris for the Sun and Moon, various parameters, constants, and the Besselian elements (polynomial form) are given in Table 1. The eclipse elements and predictions were derived from the DE200 and LE200 ephemerides (solar and lunar, respectively) developed jointly by the Jet Propulsion Laboratory and the U. S. Naval Observatory for use in the *Astronomical Almanac* for 1984 and thereafter. Unless otherwise stated, all predictions are based on center of mass positions for the Moon and Sun with no corrections made for center of figure, lunar limb profile or atmospheric refraction. The predictions depart from normal IAU convention through the use of a smaller constant for the mean lunar radius k for all umbral contacts (see: LUNAR LIMB PROFILE). Times are expressed in either Terrestrial Dynamical Time (TDT) or in Universal Time (UT), where the best value of ΔT[4] available at the time of preparation is used.

From the polynomial form of the Besselian elements, any element can be evaluated for any time 't_1' (in decimal hours) via the equation:

$$a = a_0 + a_1*t + a_2*t^2 + a_3*t^3 \quad (\text{or } a = \sum [a_n*t^n]; n = 0 \text{ to } 3)$$

where: $\mathbf{a}$ = x, y, d, l_1, l_2, or μ
$t = t_1 - t_0$ (decimal hours) and t_0 = 17.000 TDT

The polynomial Besselian elements were derived from a least-squares fit to elements rigorously calculated at five separate times over a six hour period centered at t_0. Thus, the equation and elements are valid over the period $14.00 \leq t_0 \leq 20.00$ TDT.

Table 2 lists all external and internal contacts of penumbral and umbral shadows with Earth. They include TDT times and geodetic coordinates with and without corrections for ΔT. The contacts are defined:

P1 - Instant of first external tangency of penumbral shadow cone with Earth's limb.
(partial eclipse begins)
P2 - Instant of first internal tangency of penumbral shadow cone with Earth's limb.
P3 - Instant of last internal tangency of penumbral shadow cone with Earth's limb.
P4 - Instant of last external tangency of penumbral shadow cone with Earth's limb.
(partial eclipse ends)

U1 - Instant of first external tangency of umbral shadow cone with Earth's limb.
(umbral eclipse begins)
U2 - Instant of first internal tangency of umbral shadow cone with Earth's limb.
U3 - Instant of last internal tangency of umbral shadow cone with Earth's limb.
U4 - Instant of last external tangency of umbral shadow cone with Earth's limb.
(umbral eclipse ends)

Similarly, the northern and southern extremes of the penumbral and umbral paths, and extreme limits of the umbral center line are given. The IAU longitude convention is used throughout this publication (i.e. - for longitude, east is positive and west is negative; for latitude, north is positive and south is negative).

The path of the umbral shadow is delineated at five minute intervals in Universal Time in Table 3. Coordinates of the northern limit, the southern limit and the center line are listed to the nearest tenth of an arc-minute (~185 m at the Equator). The Sun's altitude, path width and umbral duration are calculated for the center line position. Table 4 presents a physical ephemeris for the umbral shadow at five minute intervals in UT. The center line coordinates are followed by the topocentric ratio of the apparent diameters of the Moon and Sun, the eclipse obscuration[5], and the Sun's altitude and azimuth at that instant. The central path width, the umbral shadow's major and minor axes and its instantaneous velocity with respect to Earth's surface are included. Finally, the center line duration of the umbral phase is given.

Local circumstances for each center line position listed in Tables 3 and 4 are presented in Table 5. The first three columns give the Universal Time of maximum eclipse, the center line duration of totality and the altitude of the Sun at that instant. The following columns list each of the four eclipse contact times

---

[4] ΔT is the difference between Terrestrial Dynamical Time and Universal Time
[5] Eclipse obscuration is defined as the fraction of the Sun's surface area occulted by the Moon.

followed by their related contact position angles and the corresponding altitude of the Sun. The four contacts identify significant stages in the progress of the eclipse. They are defined as follows:

First Contact — Instant of first external tangency between the Moon and Sun.
(partial eclipse begins)
Second Contact — Instant of first internal tangency between the Moon and Sun.
(central or umbral eclipse begins; total or annular eclipse begins)
Third Contact — Instant of last internal tangency between the Moon and Sun.
(central or umbral eclipse ends; total or annular eclipse ends)
Fourth Contact — Instant of last external tangency between the Moon and Sun.
(partial eclipse ends)

The position angles **P** and **V** identify the point along the Sun's disk where each contact occurs[6]. Second and third contact altitudes are omitted since they are always within 1° of the altitude at maximum eclipse.

Table 6 presents topocentric values from the central path at maximum eclipse for the Moon's horizontal parallax, semi-diameter, relative angular velocity with respect to the Sun, and libration in longitude. The altitude and azimuth of the Sun are given along with the azimuth of the umbral path. The northern limit position angle identifies the point on the lunar disk defining the umbral path's northern limit. It is measured counter-clockwise from the north point of the Moon. In addition, corrections to the path limits due to the lunar limb profile are listed. The irregular profile of the Moon results in a zone of 'grazing eclipse' at each limit that is delineated by interior and exterior contacts of lunar features with the Sun's limb. This geometry is described in greater detail in the section LIMB CORRECTIONS TO THE PATH LIMITS: GRAZE ZONES. Corrections to center line durations due to the lunar limb profile are also included. When added to the durations in Tables 3, 4, 5 and 7, a slightly shorter central total phase is predicted.

To aid and assist in the plotting of the umbral path on large scale maps, the path coordinates are also tabulated at 1° intervals in longitude in Table 7. The latitude of the northern limit, southern limit and center line for each longitude is tabulated to the nearest hundredth of an arc-minute (~18.5 m at the Equator) along with the Universal Time of maximum eclipse at each position. Finally, local circumstances on the center line at maximum eclipse are listed and include the Sun's altitude and azimuth, the umbral path width and the central duration of totality.

In applications where the zones of grazing eclipse are needed in great detail, Table 8 lists these coordinates over land based portions of the path at 7.5' intervals in longitude. The time of maximum eclipse is given for both northern and southern grazing zones as well as the Sun's center line circumstances (altitude and azimuth) and the azimuth of the umbral path. The Elevation Factor and Scale Factor are also given (See: LIMB CORRECTIONS TO THE PATH LIMITS: GRAZE ZONES).

---

[6] P is defined as the contact angle measured counter-clockwise from the *north* point of the Sun's disk.
V is defined as the contact angle measured counter-clockwise from the *zenith* point of the Sun's disk.

LOCAL CIRCUMSTANCES TABLES

Local circumstances for approximately 940 cities, metropolitan areas and places in the Western Hemisphere, Europe and Africa are presented in Tables 9 through 21. In addition, local circumstances for 118 astronomical observatories listed in the *Astronomical Almanac for 1996* are given in Table 22. These tables give the local circumstances at each contact and at maximum eclipse[7] for every location. The coordinates are listed along with the location's elevation (meters) above sea-level, if known. If the elevation is unknown (i.e. - not in the data base), then the local circumstances for that location are calculated at sea-level. In any case, the elevation does not play a significant role in the predictions unless the location is near the umbral path limits and the Sun's altitude is relatively small (<10°). The Universal Time of each contact is given to the nearest second, along with position angles **P** and **V** and the altitude of the Sun. The position angles identify the point along the Sun's disk where each contact occurs and are measured counter-clockwise (i.e. - eastward) from the north and zenith points, respectively. Locations outside the umbral path miss the umbral eclipse and only witness first and fourth contacts. The Universal Time of maximum eclipse (either partial or total) is listed to the nearest second. Next, the position angles **P** and **V** of the Moon's disk with respect to the Sun are given, followed by the altitude and azimuth of the Sun at maximum eclipse. Finally, the corresponding eclipse magnitude and obscuration are listed. For umbral eclipses (both annular and total), the eclipse magnitude is identical to the topocentric ratio of the Moon's and Sun's apparent diameters. The eclipse magnitude is always less than 1 for annular eclipses and equal to or greater than 1 for total eclipses. The final column gives the duration of totality if this location lies in the path of the Moon's umbral shadow. The effects of refraction have not been included in these calculations, nor have there been any corrections for center of figure or the lunar limb profile.

Locations were chosen based on general geographic distribution, population, and proximity to the path. The primary source for geographic coordinates is *The New International Atlas* (Rand McNally, 1991). Elevations for major cities were taken from *Climates of the World* (U. S. Dept. of Commerce, 1972). The coordinates for astronomical observatories are from the *Astronomical Almanac for 1996*. In this rapidly changing political world, it is often difficult to ascertain the correct name or spelling for a given location. Therefore, the information presented here is for location purposes only and is not meant to be authoritative. Furthermore, it does not imply recognition of status of any location by the United States Government. Corrections to names, spellings, coordinates and elevations is solicited in order to update the geographic data base for future eclipse predictions.

DETAILED MAPS OF THE UMBRAL PATH

In previous NASA eclipse bulletins, the umbral path was plotted on maps from the Defense Mapping Agency's GNC (Global Navigation and Planning Charts) series which have a scale of 1:5,000,000 (1 centimeter = 50 kilometers). For the current bulletin, we have elected to use maps from the ONC (Operational Navigation Charts) series. The ONC's higher scale of 1:1,000,000 (1 centimeter = 10 kilometers) provides better coverage of major cities, highways, airports, rivers, bodies of water and basic topography required for eclipse expedition planning including site selection, transportation logistics and weather contingency strategies. Charts of the ONC series use the Lambert conformal conic projection.

The path of totality has been plotted on the following ONC charts:

ONC	M-24	(Galápagos Islands)
ONC	L-26	(Colombia)
ONC	K-26	(Panama, Colombia, Venezuela)
ONC	K-27	(Venezuela, Caribbean Sea)
ONC	J-27	(Caribbean Sea)

Sections of the above charts where the path crosses land have been selected and divided into 10 maps appearing in the last section of this publication. Northern and southern limits as well as the center line of the path are plotted using data from Table 7. Although no corrections have been made for center of figure or lunar limb profile, they have little or no effect at this scale. Atmospheric refraction has not been

---

[7] For partial eclipses, maximum eclipse is the instant when the greatest fraction of the Sun's diameter is occulted. For umbral eclipses (total or annular), maximum eclipse is the instant of mid-totality or mid-annularity.

included as its effects play a significant role only at very low solar altitudes. In any case, refraction corrections to the path are uncertain since they depend on the atmospheric temperature-pressure profile, which cannot be predicted in advance. If observations from the graze zones are planned, then the zones of grazing eclipse must be plotted on higher scale maps using coordinates in Table 8. See PLOTTING THE PATH ON MAPS for sources and more information. The ONC paths also depict the curve of maximum eclipse at five minute increments in Universal Time from Table 3.

ESTIMATING TIMES OF SECOND AND THIRD CONTACTS

The times of second and third contact for any location not listed in this publication can be estimated using the detailed maps found in the final section. Alternatively, the contact times can be estimated from maps on which the umbral path has been plotted. Table 7 lists the path coordinates conveniently arranged in 1° increments of longitude to assist plotting by hand. The path coordinates in Table 3 define a line of maximum eclipse at five minute increments in time. These lines of maximum eclipse each represent the projection diameter of the umbral shadow at the given time. Thus, any point on one of these lines will witness maximum eclipse (i.e.: mid-totality) at the same instant. The coordinates in Table 3 should be added to the map in order to construct lines of maximum eclipse.

The estimation of contact times for any one point begins with an interpolation for the time of maximum eclipse at that location. The time of maximum eclipse is proportional to a point's distance between two adjacent lines of maximum eclipse, measured along a line parallel to the center line. This relationship is valid along most of the path with the exception of the extreme ends, where the shadow experiences its largest acceleration. The center line duration of totality **D** and the path width **W** are similarly interpolated from the values of the adjacent lines of maximum eclipse as listed in Table 3. Since the location of interest probably does not lie on the center line, it is useful to have an expression for calculating the duration of totality **d** as a function of its perpendicular distance **a** from the center line:

$$\mathbf{d} = \mathbf{D} \, (1 - (2 \, \mathbf{a}/\mathbf{W})^2)^{1/2} \text{ seconds} \tag{1}$$

where: **d** = duration of totality at desired location (seconds)
D = duration of totality on the center line (seconds)
a = perpendicular distance from the center line (kilometers)
W = width of the path (kilometers)

If t_m is the interpolated time of maximum eclipse for the location, then the approximate times of second and third contacts (t_2 and t_3, respectively) are:

Second Contact: $t_2 = t_m - d/2$ [2]
Third Contact: $t_3 = t_m + d/2$ [3]

The position angles of second and third contact (either **P** or **V**) for any location off the center line are also useful in some applications. First, linearly interpolate the center line position angles of second and third contacts from the values of the adjacent lines of maximum eclipse as listed in Table 5. If X_2 and X_3 are the interpolated center line position angles of second and third contacts, then the position angles x_2 and x_3 of those contacts for an observer located **a** kilometers from the center line are:

Second Contact: $x_2 = X_2 - \arcsin(2\,\mathbf{a}/\mathbf{W})$ [4]
Third Contact: $x_3 = X_3 + \arcsin(2\,\mathbf{a}/\mathbf{W})$ [5]

where: x_n = interpolated position angle (either **P** or **V**) of contact **n** at location
X_n = interpolated position angle (either **P** or **V**) of contact **n** on center line
a = perpendicular distance from the center line (kilometers)
     (use negative values for locations south of the center line)
W = width of the path (kilometers)

MEAN LUNAR RADIUS

A fundamental parameter used in eclipse predictions is the Moon's radius k, expressed in units of Earth's equatorial radius. The Moon's actual radius varies as a function of position angle and libration due to the irregularity in the limb profile. From 1968 through 1980, the Nautical Almanac Office used two separate values for k in their predictions. The larger value (k=0.2724880), representing a mean over topographic features, was used for all penumbral (exterior) contacts and for annular eclipses. A smaller value (k=0.272281), representing a mean minimum radius, was reserved exclusively for umbral (interior) contact calculations of total eclipses [*Explanatory Supplement*, 1974]. Unfortunately, the use of two different values of k for umbral eclipses introduces a discontinuity in the case of hybrid or annular-total eclipses.

In August 1982, the International Astronomical Union (IAU) General Assembly adopted a value of k=0.2725076 for the mean lunar radius. This value is now used by the Nautical Almanac Office for all solar eclipse predictions [Fiala and Lukac, 1983] and is currently the best mean radius, averaging mountain peaks and low valleys along the Moon's rugged limb. The adoption of one single value for k is eliminates the discontinuity in the case of annular-total eclipses and ends confusion arising from the use of two different values. However, the use of even the best 'mean' value for the Moon's radius introduces a problem in predicting the true character and duration of umbral eclipses, particularly total eclipses. A total eclipse can be defined as an eclipse in which the Sun's disk is completely occulted by the Moon. This cannot occur so long as any photospheric rays are visible through deep valleys along the Moon's limb [Meeus, Grosjean and Vanderleen, 1966]. But the use of the IAU's mean k guarantees that some annular or annular-total eclipses will be misidentified as total. A case in point is the eclipse of 3 October 1986. Using the IAU value for k, the *Astronomical Almanac* identified this event as a total eclipse of 3 seconds duration when it was, in fact, a beaded annular eclipse. Since a smaller value of k is more representative of the deeper lunar valleys and hence the minimum solid disk radius, it helps ensure the correct identification of an eclipse's true nature.

Of primary interest to most observers are the times when umbral eclipse begins and ends (second and third contacts, respectively) and the duration of the umbral phase. When the IAU's value for k is used to calculate these times, they must be corrected to accommodate low valleys (total) or high mountains (annular) along the Moon's limb. The calculation of these corrections is not trivial but must be performed, especially if one plans to observe near the path limits [Herald, 1983]. For observers near the center line of a total eclipse, the limb corrections can be more closely approximated by using a smaller value of k which accounts for the valleys along the profile.

This publication uses the IAU's accepted value of k=0.2725076 for all penumbral (exterior) contacts. In order to avoid eclipse type misidentification and to predict central durations which are closer to the actual durations at total eclipses, we depart from standard convention by adopting the smaller value of k=0.272281 for all umbral (interior) contacts. This is consistent with predictions in *Fifty Year Canon of Solar Eclipses: 1986 - 2035* [Espenak, 1987]. Consequently, the smaller k produces shorter umbral durations and narrower paths for total eclipses when compared with calculations using the IAU value for k. Similarly, predictions using a smaller k result in longer umbral durations and wider paths for annular eclipses than do predictions using the IAU's k.

LUNAR LIMB PROFILE

Eclipse contact times, magnitude and duration of totality all depend on the angular diameters and relative velocities of the Moon and Sun. Unfortunately, these calculations are limited in accuracy by the departure of the Moon's limb from a perfectly circular figure. The Moon's surface exhibits a rather dramatic topography, which manifests itself as an irregular limb when seen in profile. Most eclipse calculations assume some mean radius that averages high mountain peaks and low valleys along the Moon's rugged limb. Such an approximation is acceptable for many applications, but if higher accuracy is needed, the Moon's actual limb profile must be considered. Fortunately, an extensive body of knowledge exists on this subject in the form of Watts' limb charts [Watts, 1963]. These data are the product of a photographic survey of the marginal zone of the Moon and give limb profile heights with respect to an adopted smooth reference surface (or datum). Analyses of lunar occultations of stars by Van Flandern [1970] and Morrison [1979] have shown that the average cross-section of Watts' datum is slightly elliptical rather than circular. Furthermore, the implicit center of the datum (i.e. - the center of figure) is displaced from the Moon's center of mass. In a follow-up analysis of 66,000 occultations, Morrison and Appleby [1981] have found that the radius of the datum appears to vary with libration. These variations produce systematic errors in Watts' original limb profile heights that attain 0.4 arc-seconds at some position angles. Thus, corrections to Watts' limb data are necessary to ensure that the reference datum is a sphere with its center at the center of mass.

The Watts charts have been digitized by Her Majesty's Nautical Almanac Office in Herstmonceux, England, and transformed to grid-profile format at the U. S. Naval Observatory. In this computer readable form, the Watts limb charts lend themselves to the generation of limb profiles for any lunar libration. Ellipticity and libration corrections may be applied to refer the profile to the Moon's center of mass. Such a profile can then be used to correct eclipse predictions which have been generated using a mean lunar limb.

Along the 1998 eclipse path, the Moon's topocentric libration (physical + optical) in longitude ranges from $l=-1.7°$ to $l=-3.4°$. Thus, a limb profile with the appropriate libration is required in any detailed analysis of contact times, central durations, etc.. Since the land based portions of the path occur over a narrow range of librations (i.e.: $l=-2.3°$ to $l=-3.1°$), a profile with an intermediate value is useful for general planning purposes and may even be adequate for most applications. The lunar limb profile presented in Figure 14 includes corrections for center of mass and ellipticity [Morrison and Appleby, 1981]. It is generated for 18:00 UT, which corresponds to northern Colombia near Venezuela. The Moon's topocentric libration is $l=-2.81°$, and the topocentric semi-diameters of the Sun and Moon are 969.1 and 1011.1 arc-seconds, respectively. The Moon's angular velocity with respect to the Sun is 0.358 arc-seconds per second.

The radial scale of the limb profile in Figure 14 (at bottom) is greatly exaggerated so that the true limb's departure from the mean lunar limb is readily apparent. The mean limb with respect to the center of figure of Watts' original data is shown (dashed) along with the mean limb with respect to the center of mass (solid). Note that all the predictions presented in this publication are calculated with respect to the latter limb unless otherwise noted. Position angles of various lunar features can be read using the protractor marks along the Moon's mean limb (center of mass). The position angles of all four contact points are clearly marked along with the north pole of the Moon's axis of rotation and the observer's zenith at mid-totality. The dashed line with arrows at either end identifies the contact points on the limb corresponding to the northern and southern limits of the path. To the upper left of the profile are the Sun's topocentric coordinates at maximum eclipse. They include the right ascension **R.A.**, declination **Dec.**, semi-diameter **S.D.** and horizontal parallax **H.P.**. The corresponding topocentric coordinates for the Moon are to the upper right. Below and left of the profile are the geographic coordinates of the center line at 18:00 UT while the times of the four eclipse contacts at that location appear to the lower right. Directly below the profile are the local circumstances at maximum eclipse. They include the Sun's altitude and azimuth, the path width, and central duration. The position angle of the path's northern/southern limit axis is **PA(N.Limit)** and the angular velocity of the Moon with respect to the Sun is **A.Vel.(M:S)**. At the bottom left are a number of parameters used in the predictions, and the topocentric lunar librations appear at the lower right.

In investigations where accurate contact times are needed, the lunar limb profile can be used to correct the nominal or mean limb predictions. For any given position angle, there will be a high mountain (annular eclipses) or a low valley (total eclipses) in the vicinity that ultimately determines the true instant of contact. The difference, in time, between the Sun's position when tangent to the contact point on the mean limb and tangent to the highest mountain (annular) or lowest valley (total) at actual contact is the desired correction to the predicted contact time. On the exaggerated radial scale of Figure 14, the Sun's limb can be represented as an epicyclic curve that is tangent to the mean lunar limb at the point of contact and departs from the limb by **h** through:

$$\mathbf{h} = \mathbf{S}\,(\mathbf{m}-1)\,(1-\cos[\mathbf{C}]) \qquad [6]$$

where: **h** = departure of Sun's limb from mean lunar limb
S = Sun's semi-diameter
m = eclipse magnitude
C = angle from the point of contact

Herald [1983] has taken advantage of this geometry to develop a graphical procedure for estimating correction times over a range of position angles. Briefly, a displacement curve of the Sun's limb is constructed on a transparent overlay by way of equation [6]. For a given position angle, the solar limb overlay is moved radially from the mean lunar limb contact point until it is tangent to the lowest lunar profile feature in the vicinity. The solar limb's distance **d** (arc-seconds) from the mean lunar limb is then converted to a time correction Δ by:

$$\Delta = \mathbf{d}\,\mathbf{v}\,\cos[\mathbf{X} - \mathbf{C}] \qquad [7]$$

where: Δ = correction to contact time (seconds)

d = distance of Solar limb from Moon's mean limb (arc-sec)
v = angular velocity of the Moon with respect to the Sun (arc-sec/sec)
X = center line position angle of the contact
C = angle from the point of contact

This operation may be used for predicting the formation and location of Baily's beads. When calculations are performed over a large range of position angles, a contact time correction curve can then be constructed.

Since the limb profile data are available in digital form, an analytical solution to the problem is possible that is quite straightforward and robust. Curves of corrections to the times of second and third contact for most position angles have been computer generated and are plotted in Figure 14. The circular protractor scale at the center represents the nominal contact time using a mean lunar limb. The departure of the contact correction curves from this scale graphically illustrates the time correction to the mean predictions for any position angle as a result of the Moon's true limb profile. Time corrections external to the circular scale are added to the mean contact time; time corrections internal to the protractor are subtracted from the mean contact time. The magnitude of the time correction at a given position angle is measured using any of the four radial scales plotted at each cardinal point.

For example, table 17 gives the following data for Maracaibo, Venezuela:

Second Contact = 18:04:03 UT $P_2=103°$
Third Contact = 18:06:53 UT $P_3=198°$

Using Figure 14, the measured time corrections and the resulting contact times (to the nearest second) are:

C_2=−2.5 seconds; Second Contact = 18:04:03 −2s = 18:04:01 UT
C_3=−1.0 seconds; Third Contact = 18:06:53 −1s = 18:06:52 UT

LIMB CORRECTIONS TO THE PATH LIMITS: GRAZE ZONES

The northern and southern umbral limits provided in this publication were derived using the Moon's center of mass and a mean lunar radius. They have not been corrected for the Moon's center of figure or the effects of the lunar limb profile. In applications where precise limits are required, Watts' limb data must be used to correct the nominal or mean path. Unfortunately, a single correction at each limit is not possible since the Moon's libration in longitude and the contact points of the limits along the Moon's limb each vary as a function of time and position along the umbral path. This makes it necessary to calculate a unique correction to the limits at each point along the path. Furthermore, the northern and southern limits of the umbral path are actually paralleled by a relatively narrow zone where the eclipse is neither penumbral nor umbral. An observer positioned here will witness a slender solar crescent that is fragmented into a series of bright beads and short segments whose morphology changes quickly with the rapidly varying geometry of the Moon with respect to the Sun. These beading phenomena are caused by the appearance of photospheric rays that alternately pass through deep lunar valleys and hide behind high mountain peaks as the Moon's irregular limb grazes the edge of the Sun's disk. The geometry is directly analogous to the case of grazing occultations of stars by the Moon. The graze zone is typically five to ten kilometers wide and its interior and exterior boundaries can be predicted using the lunar limb profile. The interior boundaries define the actual limits of the umbral eclipse (both total and annular) while the exterior boundaries set the outer limits of the grazing eclipse zone.

Table 6 provides topocentric data and corrections to the path limits due to the true lunar limb profile. At five minute intervals, the table lists the Moon's topocentric horizontal parallax, semi-diameter, relative angular velocity of the Moon with respect to the Sun and lunar libration in longitude. The Sun's center line altitude and azimuth is given, followed by the azimuth of the umbral path. The position angle of the point on the Moon's limb which defines the northern limit of the path is measured counter-clockwise (i.e. - eastward) from the north point on the limb. The path corrections to the northern and southern limits are listed as interior and exterior components in order to define the graze zone. Positive corrections are in the northern sense while negative shifts are in the southern sense. These corrections (minutes of arc in latitude) may be added directly to the path coordinates listed in Table 3. Corrections to the center line umbral durations due to the lunar limb profile are also included and they are all negative. Thus, when added to the central durations given in Tables 3, 4, 5 and 7, a slightly shorter central total phase is predicted.

Table 8 directly tabulates the coordinates of the zones of grazing eclipse at each limit along all land based sections of the path. The coordinates are given every 7.5' in longitude and include the time of maximum eclipse in the northern and southern graze zones as well as on the center line. The Sun's center line position (altitude and azimuth) and the path's azimuth are also listed. The Elevation Factor is the factor by which the path must be shifted north perpendicular to itself for each unit of elevation (height) above sea level. In practice, a location's elevation in multiplied by the Elevation Factor to obtain the shift. Negative values (usually the case for eclipses in the Northern Hemisphere) indicate that the path must be shifted south for positive elevations. For instance, if one's elevation is 1000 meters above sea level and the Elevation Factor is –0.20, then the shift is 1000m x –0.20 = –200m. Thus, the observer must shift the path coordinates 200 meters in a direction perpendicular to the path and in a negative or southerly sense.

The final column of Table 8 lists the Scale Factor (km/arc-second). This parameter provides an indication of the width of the zone of grazing phenomena, due to the topocentric distance of the Moon and the projection geometry of the Moon's shadow on Earth's surface. For example, let us assume a value of 2 km/arc-seconds for the Scale Factor. Since the chromosphere has an apparent thickness of about 3 arc-seconds, it would then be visible continuously during totality for any observer in the path who is within 6 kilometers of each interior limit. The most dynamic beading phenomena occurs within 1.5 arc-seconds of the Moon's limb. Using the above Scale Factor, this projects to the first 3 kilometers inside the interior limits. However, observers should position themselves at least 1 kilometer inside the interior limits (south of the northern interior limit or north of the southern interior limit) to ensure that one is inside the path due to small uncertainties in Watts' limb datum and the actual path limits.

SAROS HISTORY

The total eclipse of 1998 February 26 is the fifty-first member of saros series 130 (Table 23), as defined by van den Bergh [1955]. All eclipses in the series occur at the Moon's descending node and gamma[8] increases with each member in the family. The series is a mature one which began with a modest partial eclipse at high southern hemisphere latitudes on 1096 Aug 20. After twenty-one partial eclipses in the series and nearly three centuries, the first umbral eclipse occurred on 1475 Mar 25. The event was a two minute total eclipse through the South Pacific. During the next one and a half centuries, the umbral duration continued to increase as each path shifted progressively northward. The greatest umbral duration of Saros 130 occurred during the total eclipse of 1619 Jul 11. Unfortunately, the 6 minute 41 second total eclipse was only visible from equatorial Africa which was virtually inaccessible to astronomers of the day.

As the duration of each succeeding eclipse decreased, the paths reversed their northern migration and drifted southward during the 18th and 19th centuries. This effect occurred as a result of Earth's passage through winter solstice whereby the northern hemisphere tipped away from the Sun. A notable member of the series occurred on 1871 Dec 12. Spectroscopic observations of this event made by French astronomer Pierre Jules Janssen led him to propose that the corona is a physical part of the Sun and is composed of both hot gases and cooler particles. Together with observations of a later eclipse in 1778, it convinced Janssen that the shape of the corona is linked to the sunspot cycle.

The northbound trend of the Saros series resumed with the eclipse of 1908 Jan 3. At this point, the duration of totality at greatest eclipse had dropped to 4 minutes 14 seconds. The most recent member occurred on 1980 February 20 and its path crossed East Africa, the Indian Ocean and central India. After 1998, the next member occurs on 2016 Mar 9 and passes through Indonesia and the Pacific. The total eclipse of 2034 Mar 20 swings through central Africa, the Middle East and South Asia. Three saros cycles after 1998, the series returns to the western hemisphere producing a path passing through central Mexico and the southeastern United States. The duration of totality drops as Saros 130 continues to produce total eclipses during the 22nd century. The last umbral eclipse of the series occurs on 2232 Jul 18 and lasts a maximum of 1 minute 14 seconds. The final nine eclipses of the series are all partial events in the polar regions of the northern hemisphere. The family terminates with the partial eclipse of 2394 Oct 25. A detailed list of eclipses in saros series 130 appears in Table 23.

In summary, Saros series 130 includes 73 eclipses with the following distribution:

Saros 130	*Partial*	*Annular*	*Ann/Total*	*Total*
Non-Central	30	0	0	0
Central	—	0	0	43

---

[8] Minimum distance of the Moon's shadow axis from Earth's center in units of equatorial Earth radii. Gamma defines the instant of greatest eclipse and takes on negative values south of the Earth's center.

Weather Prospects for the Eclipse

Overview

After the frigid winter eclipse of 1997, Nature now treats us to a sub-tropical extravaganza during the final months of the northern hemisphere winter. While the land-based choices are relatively limited for this track (in comparison with that upcoming in 1999), the weather prospects are generally very favorable.

The eclipse track passes over the Pacific's equatorial doldrums, skipping through the Galápagos before landing on the Pacific shores of Panama and Colombia. Even this area, one of the wettest in the world, is in the midst of its dry season, with subdued cloudiness and rainfall. Over the Andes into northern Colombia and Venezuela the track enters a very dry, sunny climate with cooling trade winds. February is the dry season in this already desiccated area, so the prospects for sunny skies are very good. Farther east, over the Leeward Islands, cloud cover increases slightly, but fine weather persists into the mid Atlantic.

The Galápagos Islands

The first landing of the Moon's shadow is over the legendary Galápagos Islands of Ecuador (officially known as the Archipelago de Colon), lying astride the equator, about 400 km west of the South American coast. The islands are the scattered rocky promontories of oceanic volcanoes, bathed in the flow of the Equatorial and Humboldt currents. The prevailing easterly winds, cold currents, and equatorial location combine to bring a variable climatology and an exotic ecology.

The eclipse track passes over the northern portion of the archipelago, with a center line duration of 3m 59s and a Sun altitude of 69°. Unfortunately, the center line misses all opportunities to make landfall. The island closest to the center line is Isla Pinta, a small volcanic peak protruding 750 meters above the waves. The island is uninhabited, though not always unoccupied. It is the offspring of two volcanoes, one old and worn, the other newer, but no eruption has been recorded on the island. Access is difficult, usually by chartered boat from the settlements of the larger islands to the south. Depending on the state of the sea, landing on Pinta may be wet, and equipment should be well protected. A licensed guide is required.

Several other islands also fall within the eclipse track, primarily the northern tip of the largest island, Isla Isabela, a small piece of Isla Fernandina, and Isla Marchena. All of the locations lying under the eclipse track are zoned for scientific use, a designation described by Constance (1995) as having "no possible access to the casual visitor with no specialized interest." Presumably the eclipse would constitute such a specialized interest. But in order to facilitate compliance with zoning regulations, eclipse travelers who wish to view from the Galápagos are likely best advised to travel as part of an organized group, or arrive early to arrange charters and permission to travel into the umbral zone.

Weather on the Galápagos is dictated by the winds and ocean currents, but this simple relationship is complicated by the archipelago's location at the confluence of two currents and two wind streams, so that the controlling forces change from season to season. Oceanic influences vary between the cool Humboldt Current from the coast of Peru, and the warmer Panama Flow which arrives from the northeast. Winds blow steadily from the southeast quadrant 67% of the time, but the islands lie only a short distance south of the Intertropical Convergence Zone (ICZ), a region of enhanced convective cloudiness caused by the collision of northerly winds from the northern hemisphere and southerlies from the south.

During February the ICZ is closest to the islands, and the wettest season is in full sway (the peak is in March). Observations at Puerto Baquerizo on Isla Cristobal show rainfall on nearly a third of the days of February (Table 24). Fortunately, February is also one of the sunniest months, with an average of 7 1/2 hours of sunshine brightening each day at Puerto Baquerizo and Darwin Station (at Puerto Ayora on Isla Santa Cruz). This contradiction between sunshine and precipitation is caused by the nature of the clouds, which tend to be convective, bringing brief periods of heavy showers between generous doses of sunshine. In fact Puerto Baquerizo is illuminated by 63% of the maximum sunshine amount possible, a number which also closely describes the probability of seeing the eclipse. Some minor adjustment in this probability is required for the time of day of the eclipse (likely a small increase), the location of the track (a slight downward adjustment for the more northerly islands) and the fact that "percent of possible sunshine" is calculated assuming a flat horizon with no refraction.

Convective buildups tend to come in the afternoon, promoted by upslope winds on the south and east sides of the islands. This suggests that the best eclipse sites will lie in the lee of the peaks, generally on the north and west sides. There is no guarantee here however, as cloud tops from the upwind side are easily blown downwind over the eclipse site and the weather may turn grey rather quickly if the day is particularly unstable. In fact the complicated wind flow over and around the volcanic terrain may bring showers at any location wherever two flows combine. The presence and dominance of cacti among the local vegetation will quickly identify the drier slopes and the most promising eclipse sites.

Mobility would be an advantage, but is probably not a viable option in view of the difficult terrain and lack of roads. Eclipse viewing from sea may offer a little more mobility, and a spot in the lee of the land will bring smaller waves than the one meter seas which are found in nearby open waters (Figure 16).

Satellite-derived cloud statistics (Figure 15) show that the mean cloud cover over the islands is a respectable 45%. This figure is an average over a 5° latitude by 5° longitude box, and includes a substantial area of ocean, thus giving a relatively coarse measurement of the highly variable cloudiness over the islands. While this is less illuminating than sunshine measurements from the ground, it has the advantage of making statistics more comparable from country to country, since it is determined in the same fashion at each location on the globe. Figure 15 and Table 24 indicate that sites in the Galápagos, though relatively sunny, are not as good as those over northern South America.

SOUTH AMERICA

Between the Galápagos Islands and the South American continent, the shadow track moves across the ICZ and mean cloudiness increases to a maximum of 60%. Ordinarily this is one of the cloudiest areas of the globe, but the thunderstorm activity on the ICZ is very much subdued in February and dry weather is the rule rather than the exception. The influence of the ICZ extends onto the coast of Colombia, helping to create one of the world's wettest habitats, but the eclipse path fortunately skirts a little to the north of this sopping region. The mean cloud cover at the Panamanian coast is nearly the same as over the Galápagos.

Travel along the Panama-Columbia border is quite difficult. Heavy rains and a swampy landscape have conspired to foil completion of the Pan-American highway through this region, though atlases show a rudimentary road connection snaking from Panama City to the Colombian Pacific coast. Travel is likely best by boat to the vicinity of Jaqué, just north of the center line on the coast. Weather statistics for Jaqué are limited, but they show that mean rainfall is only 3 mm in February, compared to over 950 mm in October. The eclipse could not occur at a better time, as February is the driest month of the year here.

Winds in the Gulf of Panama have no preferred direction, but northeasterlies begin to dominate close to and over the Isthmus of Panama. This implies that Pacific coast sites may be drier and sunnier than those on the Caribbean side, as the air is flowing down the western side of the mountainous backbone of Panama. Air which is descending is drier than that which is flowing uphill. Wave heights are also relatively low here, as the variable winds and proximity of land limit the energy imparted to the ocean surface.

Past the Isthmus of Panama, the eclipse track crosses over the Golfo de Urabá, where Central and South America come together on the Caribbean coast, and then moves inland into Colombia. Though the Andes Mountains are greatly diminished in size at this point, the prevailing northerly upslope winds, and the nearby location of a weak ICZ combine to bring a region of slightly greater cloudiness and rainfall. Turbo, on the Golfo de Urabá, reports 6 hours of sunshine per day in February (Table 24). This region is the cloudiest land based site along the entire path.

East of the Golfo de Urabá, the eclipse track crosses onto the Caribbean lowlands of Colombia. These lowlands lie between two branches of the Andes mountains which together provide some degree of protection from weather systems approaching from the northeast. Satellite measurements (Figure 15) show that cloud cover drops rapidly through this region, falling by nearly 20% from west to east across northern Colombia. Sunshine hours increase in a complementary fashion, rising from 6 hours per day at Turbo to 9 hours near Valledupar, close to the Venezuelan border. This is a very generous 80 percent of the maximum sunshine possible and eastern Colombia is an excellent location from which to observe the eclipse. Satellite measurements show less than 40% mean cloud cover here - as good as anywhere else along the path.

As the eclipse track leaves Colombia, it passes south of the 5775 m Sierra Nevada de Santa Maria. This mountain is an imposing peak, thrusting dramatically upward from sea level with a vertical rise which is comparable to the Himalayas. The eclipse track runs just south of the mountain; the north limit barely touches the summit. Just east of this solitary peak is the main branch of the Andes, the Sierra de Perijá which forms the border between Colombia and Venezuela. The valley between the Sierra Nevada and the Sierra de Perijá contains the city of Valledupar, a short distance north of the center line. The blocking effect of mountains to the east and west of the city serve to protect this segment of the track from errant weather systems which might bring cloud cover. In view of the excellent climatology, nearby track, and convenient road access, Valledupar provides a convenient base for an eclipse expedition.

Beyond Valledupar, the track climbs the Sierra de Perijá and moves into Venezuela and Lake Maracaibo. Good weather continues through the Maracaibo lowlands and out across the Gulf of Venezuela to Aruba and Curaçao. This part of northern South America - across northern Colombia and northwest

Venezuela, and continuing across the adjacent Caribbean - has an anomalously dry and sunny climate. The explanation for this semi-desert climate has eluded complete explanation, though it seems to be related to the deflection of the oceanic trade winds by the mountains along the north coast, and by the presence of a high pressure ridge extending from Bermuda. These two features cause the atmosphere to subside, drying and warming as it does, and suppressing the development of convective showers and thundershowers.

Whatever the explanation, eclipse observers will derive considerable benefit from this climatic bonus. And adding to this promising situation is the fact that in most locations along this part of the track, February is the driest or second driest month of the year, with as little as 1 mm of monthly rainfall recorded at one station (Uribia).

Still, northern South America is not completely without major winter weather systems. Rainfall statistics shows that there are uncommon instances of very heavy precipitation - more than 60 mm in 24 hours at Maracaibo for instance. These wet episodes are brought on by northern hemisphere disturbances which manage to reach low latitudes, either as cold fronts or as cold pools of air. The months of February and March are favored for such disturbances in Venezuela; in March they are known as the "invierno de las chicharras." The break in the normally dry weather is impressive, but the heaviest precipitation is reserved for the slopes of the coastal mountains where the air is forced to rise, adding to the processes which create rainfall. During such periods, cloud cover is widespread and solid, with depressing prospects for the eclipse chaser. Luckily they occur at intervals of several years - the rainfall record suggests about once every seven.

The largest city within the eclipse track is Maracaibo in Venezuela, solidly located within the best weather prospects. While the city lies south of the center line, the best spots on the track can be reached on the highway leading northward along the coast. The prevailing easterly winds off of the Gulf of Venezuela will stabilize the atmosphere and present a steadier view of the Sun, though the cooling which normally accompanies an oncoming eclipse will do this without further help.

Leaving the Maracaibo area, the track heads out across the Gulf of Venezuela, crossing onto land again on the narrow-necked Peninsula de Paraguaná. This is the last land-based location which gives access to the center line, though other island sites in the Caribbean offer excellent opportunities at a small cost in eclipse duration. For those watching from the Peninsula de Paraguaná, the refinery town of Punto Fijo would seem to be the best base, with a short trip northward required to reach the middle of the track. The eclipse duration at Punto Fijo is 3m 27s, only 20 seconds less than on the center line.

ARUBA AND CURAÇAO

North and east of the Peninsula de Paraguaná are the former Dutch islands of Aruba and Curaçao, both popular tourist destinations. For this eclipse, they share in the good weather prospects characteristic of northern South America. The eclipse center line passes between the two, so observations from these islands will impose a time penalty. At the northern tip of Curaçao, the eclipse duration of 3m 32s is only 11 seconds shorter than on the center line (3m 43s). At the southern tip of Aruba (3m 34s), the time penalty is only 9 seconds. Thus, both islands make excellent locations since they share similar weather conditions.

Aruba is a 31 by 10 kilometer island. Most hotels there are located along the northwest coast where totality is under three minutes. However, a 25 km drive to the island's southern end increases the duration by 30 seconds. Aruba's dry weather is reflected in scant vegetation - mostly cacti and wind-bent trees. February is the cool season with temperatures in the mid twenties Centigrade. Precipitation is low, approaching a minimum in March and April. February is also the sunniest month, with a mean cloud cover of less than 40%. Relief on the island is low, only 188 meters at the highest point, providing little disruption of the trade winds. The combination of dry interior and the steady trade winds brings lots of dust.

Sint Nicholas is the site of a large oil refinery, with appropriate smells and sights. At one time this was the largest refinery in the world, but has been operated at only part of its capacity since 1991. South of the refinery, along the coastal road at Seroe Colorado, is the former Exxon staff residential area (known as "the colony") which includes two west-facing beaches (Baby Beach and Rodgers Beach). According to one guide book (Cameron and Box, 1995), the residential site and beaches are accessible to the public. From an astronomical point of view the beaches offer the best location on the island , but are rather exposed to the unceasing northeast trade winds. This could be an advantage as an exposed beach (that is, exposed to the onshore trade winds) is more likely to be free of the dust which bothers the rest of the island. Colorado Point, a short distance from Seroe Colorado, may provide additional viewing sites, though it is also exposed to the winds.

The weekend before the eclipse is the end of Carnival across the Caribbean, marked by parades and other festivities. Some of the celebrations may go on into eclipse week, but sober civilization should have returned before the Moon begins its own ceremony with the Sun.

Curaçao is the largest of the islands of the Netherlands Antilles, lying 60 kilometers off the Venezuelan coast. Like Aruba, it is dry and sparsely vegetated, though with slightly more relief to block the winds. Best eclipse observing (astronomically) is on the rugged north coast. Those wishing to sacrifice a few seconds of totality may find the protected beaches and coves of the west side of the island more to their liking. Some beaches are private and offer a few amenities not available at public beaches in exchange for a user fee. Dorp Lagun is one such scenic spot on the northwest coast, not far from the northern tip of the island, offering "a lovely secluded beach in a small cove with cliffs surrounding it and small fishing boats" (Cameron and Box, 1995). Knip Bay, a little to the north of Dorp Lagun, has a larger beach.

There is no particular reason to watch this eclipse from a beach-front site since the eclipse occurs well overhead and a clear horizon is not a prerequisite, except possibly for an unobstructed view of the oncoming shadow cone. Ordinarily beach sites are preferred for the stability and sunshine promoted by cooling winds off the water, but the steady trade winds blowing over these small islands are able to readily achieve this. Stable conditions in the lower atmosphere may reduce the possibility of seeing shadow bands, but these ghostly heralds are also promoted by turbulent conditions at higher levels, and may still delight the careful observer. The western beaches offer some protection from the easterly trade winds, but similar spots might also be found in the lee of steeper hills. The highest point of the island, Mount Christoffel, east of Dorp Lagun, rises only 385 meters. A little scouting in the days before the eclipse will ferret out the best spots, according to the type of observing and measurement that you may wish to do. Those with a penchant for hiking and a small amount of equipment may wish to hike to the top of Mount Christoffel to view the eclipse - a distance of 7 to 12 km depending on the trail taken.

Sunshine statistics for Curaçao and Aruba are the equal of those in nearby Venezuela and Colombia - between 8 and 9 hours per day, or 70 to 80 percent or more of the maximum possible (Table 24). However, rainfall statistics show that Aruba is the drier island, by a small amount. Choices between continental locations in Venezuela or northeast Colombia and island locations on Aruba and Curaçao will have to be made on the basis of factors other than weather, for there is not much to distinguish one location from the other. Throughout this area, February is the sunniest month.

Small tropical weather systems, when they do approach, will come from the east, not the western horizon which is more familiar at North American and European latitudes. These easterly waves bring scattered showers and thundershowers, along with plenty of high level cloud. Most of these systems will be substantially weakened by the atmospheric subsidence along the north coast as they move toward the eclipse track. Charts of the occurrence of larger thunderstorm clusters show that none of them reach the umbral path, though they are rare visitors to the southern tip of Lake Maracaibo (Garcia, 1985). These clusters originate along the ICZ, which lies across southern Venezuela and Colombia, well away from the track.

It should be noted that thin high level cloudiness is endemic throughout tropical and subtropical latitudes, and observers should not expect pure blue skies at any location for this eclipse, though the continental locations are more likely to see clean skies than island sites. Interestingly, descending air in front of easterly waves does flush out the cirrus cloudiness, and those willing to take a chance by positioning themselves in front of an approaching disturbance may be rewarded with the best skies of all. This option is probably open only to those who travel aboard ship. Of course if you guess the speed of the disturbance incorrectly....

THE LEEWARD ISLANDS

The last land-based sites from which to watch the eclipse are over the Leeward Islands of Guadeloupe, Antigua and Montserrat and a few smaller outposts. Once again the center of the eclipse track misses land, passing down the middle of the Guadeloupe Passage. Cloud conditions are not so favorable as those along the north coast of South America, but fine enough to present a good opportunity to view the Sun at the critical moment.

Montserrat lies within the shadow path, on the north side of the center line. At Plymouth, the capital, eclipse duration is 2m 57s. A road heads south from Plymouth to St. Patrick's, about 3 km away to the south, and actually goes beyond that, close to the south tip of the island. The eclipse duration here is 3m 3s, or 14 seconds less than the duration on the center line. A volcano in the Soufriére Hills at the south end of island erupted in early 1996. It will add an extra challenge to eclipse observers if it remains active.

Guadeloupe, a French Region on the opposite side of the track and the Passage, is the largest of the islands in the eclipse shadow. It is really two islands, connected by a bridge across a narrow strait.

Almost all of the two islands lies within the zone of totality, but for most observers, the most northerly points of land will be the most attractive. At the northern tip of Basse-Terre (the larger and westernmost island), the eclipse duration is 3m 00s. The northern tip of Grande-Terre experiences the same duration.

The climatological record at Pointe-à-Pitre, the capital of Guadeloupe, shows that sunshine in February reaches two-thirds of the amount possible if it were clear from sunrise to sunset. This is about 15% less than over the Dutch Antilles, Venezuela and parts of Colombia, but still indicates a very sunny climate. Much of the cloudiness will be attached to the terrain, with hills and volcanoes deflecting the trade winds upward to promote the growth of convective clouds. Terrain is much more elevated on Basse-Terre where the 1465 m high volcano La Soufrière dominates the skyline. Usually cloaked in cloud, the volcano may be a spectacular perch from which to view the eclipse, and especially the incoming and outgoing shadow. The summit is quite flat, though with spurting sulfurous fumes on a yellow landscape.

Keep an eye on the peak for a few days before the eclipse to assess your chances of seeing the Sun at the critical moments. Cloud buildups are likely to be suppressed by the cooling associated with the oncoming shadow, but be careful about fog which will probably form very quickly between second and third contact on the windward slopes of the volcano, even if skies are perfectly clear beforehand. If you are not on the peak but instead elect for lower slopes, at the very least try to place yourself on the leeward (west) side of the mountain in time for totality.

According to annual rainfall statistics, the driest spot on Guadeloupe is the northeast coast of Grande-Terre, east of the community of Campàche. This area is near the closest point to the center line, the aptly named Pointe de la Grande Vigie. It should also be the sunniest location on the island, though eclipse conditions will depend on the weather of the day, rather than the accumulated wisdom of climatology. In all Caribbean islands the rainfall and cloudiness are related to topography. Annual rainfall at the top of La Soufrière is estimated to reach ten meters while that on the windward east coast is less than one.

Antigua is a little further from the center line than either Montserrat or Guadeloupe, and the maximum eclipse duration is of the order of 2m 47s. The more desirable spots on this island are found on the southeastern shores, near the community of English Harbor ("one of the world's most attractive yachting centres"). A good location here might be Shirley Heights, the site of 18th century fortifications and giving a view in all directions. Since the south coast of the island lies more or less parallel to the eclipse line, there is little to recommend one site in this area over another as far as timing is concerned. Hotels and tourist facilities available at the community of English Harbor will make it a popular destination, but road access is also convenient for less populated spots at Carlisle Bay and Old Road Bluff.

Sunshine statistics are not available for Montserrat or Antigua, but weather should be very similar to Guadeloupe. Mean cloud falls within the range of 40 to 50 percent with local statistics favoring the higher number. This may not represent actual conditions too well, since the higher elevations on these islands are often necklaced with clouds, which must be included in the amounts reported by ground observers. This will tend to make official reports appear more cloudy than is actually the case at the best coastal eclipse sites. Still, there is no doubt that cloud cover on the Leeward Islands is heavier than farther west along the South American coast, but if eclipses on sandy beaches with overhanging palm trees appeal to you, then the Leewards are a choice location.

The island of Redonda, northwest of Montserrat, also lies within the path, but is no more than an uninhabited small rocky outcrop. Carnival is celebrated in the latter part of February on Guadeloupe, with congested streets and slow moving traffic (1 km/h by one account). By eclipse day, festivities should be over, with Lent replacing the festivities of the week before. Carnival is held in July on Antigua.

CRUISING

Cruise boats will certainly be one of the favorite methods to see this eclipse, coming as it does toward the end of the northern hemisphere winter, and in a location which is relatively inexpensive to reach. A quick glance at 1 shows that the favorable cloud statistics which prevail along the South American coast also extend well offshore, decreasing by only a few percent between Aruba and the Leewards. With the mobility which comes with ship-board observing, and the ready availability of weather information, it is hard to see how the chances of seeing the eclipse could be anything less than a certainty, provided there is sufficient leeway built into the sailing schedule to allow small diversions to less cloudy spots.

Meticulous photographers however will find shipboard observing to be frustrating. The rolling deck is a difficult place to obtain long exposure photographs, or even hold a telescope on the Sun for more than a brief period. Unfortunately the steady trade winds across the Caribbean and the long fetch from the Leeward Islands to the Gulf of Venezuela combine to bring wave heights which average nearly 1.5 meters (2). Seas surrounding the Galápagos Islands are a little lower - about 1 meter on average, though expedition boats are likely to be smaller and thus more affected by the prevailing seas.

Calmer seas can be found in the lee of islands and in bays, particularly in the Gulf of Venezuela opposite Puerto Fijo. Other locations, perhaps west of Curaçao or behind the Leeward Islands, come at the expense of eclipse duration, since they would not be on the center line. These choices will likely be dictated by the itinerary of the cruise boat or yacht, and the flexibility of the schedule. There are numerous locations where a cruising expedition could drop off land parties and then heading off for a more central position to extract the maximum duration from the path.

OBSERVING THE ECLIPSE

EYE SAFETY DURING SOLAR ECLIPSES

The Sun can be viewed safely with the naked eye only during the few brief seconds or minutes of a *total* solar eclipse. Partial eclipses, annular eclipses, and the partial phases of total eclipses are *never* safe to watch without taking special precautions. Even when 99% of the Sun's surface is obscured during the partial phases of a total eclipse, the remaining photospheric crescent is intensely bright and cannot be viewed safely without eye protection [Chou, 1981; Marsh, 1982]. ***Do not attempt to observe the partial or annular phases of any eclipse with the naked eye. Failure to use appropriate filtration may result in permanent eye damage or blindness!***

Generally, the same equipment, techniques and precautions used to observe the Sun outside of eclipse are required for annular eclipses and the partial phases of total eclipses [Reynolds & Sweetsir, 1995; Pasachoff & Covington, 1993; Pasachoff & Menzel, 1992; Sherrod, 1981]. The safest and most inexpensive of these methods is by projection, in which a pinhole or small opening is used to cast the image of the Sun on a screen placed a half-meter or more beyond the opening. Projected images of the Sun may even be seen on the ground in the small openings created by interlacing fingers, or in the dappled sunlight beneath a leafy tree. Binoculars can also be used to project a magnified image of the Sun on a white card, but you must avoid the temptation of using these instruments for direct viewing.

The Sun can be viewed directly only when using filters specifically designed for this purpose. Such filters usually have a thin layer of aluminum, chromium or silver deposited on their surfaces that attenuates both visible and infrared energy. One of the most widely available filters for safe solar viewing is a number 14 welder's glass, available through welding supply outlets. More recently, aluminized mylar has become a popular, inexpensive alternative. Mylar can easily be cut with scissors and adapted to any kind of box or viewing device. A number of sources for solar filters are listed below. No filter is safe to use with any optical device (i.e. - telescope, binoculars, etc.) unless it has been specifically designed for that purpose. Experienced amateur and professional astronomers may also use one or two layers of completely exposed and fully developed black-and-white film, provided the film contains a silver emulsion. Since all developed color films lack silver, they are always unsafe for use in solar viewing.

Unsafe filters include color film, some non-silver black and white film, medical x-ray films with images on them, smoked glass, photographic neutral density filters and polarizing filters. Solar filters designed to thread into eyepieces which are often sold with inexpensive telescopes are also dangerous. They should not be used for viewing the Sun at any time since they often crack from overheating. Do not experiment with other filters unless you are certain that they are safe. Damage to the eyes comes predominantly from invisible infrared wavelengths. The fact that the Sun appears dark in a filter or that you feel no discomfort does not guarantee that your eyes are safe. Avoid all unnecessary risks. Your local planetarium or amateur astronomy club is a good source for additional information.

In spite of these precautions, the *total* phase of an eclipse can and should be viewed without any filters whatsoever. The naked eye view of totality is completely safe and is overwhelmingly awe-inspiring!

SOURCES FOR SOLAR FILTERS

The following is a brief list of sources for mylar and/or glass filters specifically designed for safe solar viewing with or without a telescope. The list is not meant to be exhaustive, but is simply a representative sample of sources for solar filters currently available in the United States. For additional sources, see advertisements in *Astronomy* and/or *Sky & Telescope* magazines. The inclusion of any source on this list does not imply an endorsement of that source by either of the authors or NASA.

- ABELexpress - Astronomy Division, 230-Y E. Main St., Carnegie, PA 15106. (412) 279-0672
- Celestron International, 2835 Columbia St., Torrance, CA 90503. (310) 328-9560
- Edwin Hirsch, 29 Lakeview Dr., Tomkins Cove, NY 10986. (914) 786-3738
- Meade Instruments Corporation, 16542 Millikan Ave., Irvine, CA 92714. (714) 756-2291
- Orion Telescope Center, 2450 17th Ave., PO Box 1158-S, Santa Cruz, CA 95061. (408) 464-0446
- Pocono Mountain Optics, R.R. 6, Box 6329, Moscow, PA 18444. (717) 842-1500
- Rainbow Symphony, Inc., 6860 Canby Ave., #120, Resenda, CA 91335 (510) 581-8266
- Roger W. Tuthill, Inc., 11 Tanglewood Lane, Mountainside, NJ 07092. (908) 232-1786

• Thousand Oaks Optical, Box 5044-289, Thousand Oaks, CA 91359.   (805) 491-3642

ECLIPSE PHOTOGRAPHY

The eclipse may be safely photographed provided that the above precautions are followed. Almost any kind of camera with manual controls can be used to capture this rare event. However, a lens with a fairly long focal length is recommended to produce as large an image of the Sun as possible. A standard 50 mm lens yields a minuscule 0.5 mm image, while a 200 mm telephoto or zoom produces a 1.9 mm image. A better choice would be one of the small, compact catadioptic or mirror lenses that have become widely available in the past ten years. The focal length of 500 mm is most common among such mirror lenses and yields a solar image of 4.6 mm. With one solar radius of corona on either side, an eclipse view during totality will cover 9.2 mm. Adding a 2x tele-converter will produce a 1000 mm focal length, which doubles the Sun's size to 9.2 mm. Focal lengths in excess of 1000 mm usually fall within the realm of amateur telescopes. If full disk photography of partial phases on 35 mm format is planned, the focal length of the optics must not exceed 2600 mm. However, since most cameras don't show the full extent of the image in their viewfinders, a more practical limit is about 2000 mm. Longer focal lengths permit photography of only a magnified portion of the Sun's disk. In order to photograph the Sun's corona during totality, the focal length should be no longer than 1500 mm to 1800 mm (for 35 mm equipment). However, a focal length of 1000 mm requires less critical framing and can capture some of the longer coronal streamers. For any particular focal length, the diameter of the Sun's image is approximately equal to the focal length divided by 109 (Table 25).

A mylar or glass solar filter must be used on the lens throughout the partial phases for both photography and safe viewing. Such filters are most easily obtained through manufacturers and dealers listed in *Sky & Telescope* and *Astronomy* magazines (see: Appendix 1). These filters typically attenuate the Sun's visible and infrared energy by a factor of 100,000. However, the actual filter factor and choice of ISO film speed will play critical roles in determining the correct photographic exposure. A low to medium speed film is recommended (ISO 50 to 100) since the Sun gives off abundant light. The easiest method for determining the correct exposure is accomplished by running a calibration test on the uneclipsed Sun. Shoot a roll of film of the mid-day Sun at a fixed aperture (f/8 to f/16) using every shutter speed between 1/1000 and 1/4 second. After the film is developed, note the best exposures and use them to photograph all the partial phases. The Sun's surface brightness remains constant throughout the eclipse, so no exposure compensation is necessary except for the narrow crescent phases which may require two more stops due to solar limb darkening. Bracketing by several stops may also be necessary if haze or clouds interferes on eclipse day.

Certainly the most spectacular and awe inspiring phase of the eclipse is totality. For a few brief minutes or seconds, the Sun's pearly white corona, red prominences and chromosphere are visible. The great challenge is to obtain a set of photographs which captures some aspect of these fleeting phenomena. The most important point to remember is that during the total phase, all solar filters *must be removed!* The corona has a surface brightness a million times fainter than the photosphere, so photographs of the corona are made without a filter. Furthermore, it is completely safe to view the totally eclipsed Sun directly with the naked eye. No filters are needed and they will only hinder your view. The average brightness of the corona varies inversely with the distance from the Sun's limb. The inner corona is far brighter than the outer corona. Thus, no one exposure can capture its the full dynamic range. The best strategy is to choose one aperture or f/number and bracket the exposures over a range of shutter speeds (i.e. - 1/1000 down to 1 second). Rehearsing this sequence is highly recommended since great excitement accompanies totality and there is little time to think.

Exposure times for various combinations of film speeds (ISO), apertures (f/number) and solar features (chromosphere, prominences, inner, middle and outer corona) are summarized in Table 26. The table was developed from eclipse photographs made by Espenak as well as from photographs published in *Sky and Telescope*. To use the table, first select the ISO film speed in the upper left column. Next, move to the right to the desired aperture or f/number for the chosen ISO. The shutter speeds in that column may be used as starting points for photographing various features and phenomena tabulated in the 'Subject' column at the far left. For example, to photograph prominences using ISO 100 at f/11, the table recommends an exposure of 1/500. Alternatively, you can calculate the recommended shutter speed using the 'Q' factors tabulated along with the exposure formula at the bottom of Table 26. Keep in mind that these exposures are based on a clear sky and a corona of average brightness. You should bracket your exposures one or more stops to take into account the actual sky conditions and the variable nature of these phenomena.

Another interesting way to photograph the eclipse is to record its various phases all on one frame. This is accomplished by using a stationary camera capable of making multiple exposures (check the camera instruction manual). Since the Sun moves through the sky at the rate of 15 degrees per hour, it slowly drifts through the field of view of any camera equipped with a normal focal length lens (i.e. - 35 to 50 mm). If the camera is oriented so that the Sun drifts along the frame's diagonal, it will take over three hours for the Sun to cross the field of a 50 mm lens. The proper camera orientation can be determined through trial and error several days before the eclipse. This will also insure that no trees or buildings obscure the camera's view during the eclipse. The Sun should be positioned along the eastern (left in the northern hemisphere) edge or corner of the viewfinder shortly before the eclipse begins. Exposures are then made throughout the eclipse at ~five minute intervals. The camera must remain perfectly rigid during this period and may be clamped to a wall or fence post since tripods are easily bumped. If you're in the path of totality, you'll want to remove the solar filter during the total phase and take a long exposure (~1 second) in order to record the corona in your sequence. The final photograph will consist of a string of Suns, each showing a different phase of the eclipse.

Finally, an eclipse effect that is easily captured with point-and-shoot or automatic cameras should not be overlooked. Use a kitchen sieve or colander and allow its shadow to fall on a piece of white cardboard placed several feet away. The holes in the utensil act like pinhole cameras and each one projects its own image of the Sun. The effect can also be duplicated by forming a small aperture with one's hands and watching the ground below. The pinhole camera effect becomes more prominent with increasing eclipse magnitude. Virtually any camera can be used to photograph the phenomenon, but automatic cameras must have their flashes turned off since this would otherwise obliterate the pinhole images.

For those who choose to photograph this eclipse from one of the many cruise ships in the path, some special comments are in order. Shipboard photography puts certain limits on the focal length and shutter speeds that can be used. It's difficult to make specific recommendations since it depends on the stability of the ship as well as wave heights encountered on eclipse day. Certainly telescopes with focal lengths of 1000 mm or more can be ruled out since their small fields of view would require the ship to remain virtually motionless during totality, and this is rather unlikely even given calm seas. A 500 mm lens might be a safe upper limit in focal length. Film choice could be determined on eclipse day by viewing the Sun through the camera lens and noting the image motion due to the rolling sea. If it's a calm day, you might try an ISO 100 film. For rougher seas, ISO 400 or more might be a better choice. Shutter speeds as slow as 1/8 or 1/4 may be tried if the conditions warrant it. Otherwise, stick with a 1/15 or 1/30 and shoot a sequence through 1/1000 second. It might be good insurance to bring a wider 200 mm lens just in case the seas are rougher than expected. As worst case scenario, Espenak photographed the 1984 total eclipse aboard a 95 foot yacht in seas of 3 feet. He had to hold on with one hand and point his 350 mm lens with the other! Even at that short focal length, it was difficult to keep the Sun in the field. However, any of the major cruise ships in the Caribbean will offer a far more stable platform than this.

For more information on eclipse photography, observations and eye safety, see FURTHER READING in the BIBLIOGRAPHY.

SKY AT TOTALITY

The total phase of an eclipse is accompanied by the onset of a rapidly darkening sky whose appearance resembles evening twilight about 30 to 40 minutes after sunset. The effect presents an excellent opportunity to view planets and bright stars in the daytime sky. Aside from the sheer novelty of it, such observations are useful in gauging the apparent sky brightness and transparency during totality. The Sun is in Pisces and all five naked eye planets as well as a number of bright stars will be above the horizon for observers within the umbral path. Figure 17 depicts the appearance of the sky during totality as seen from the center line at 18:00 UT. This corresponds to northern Colombia near the northwest border of Venezuela.

Mercury (m_V=–1.5) and Jupiter (m_V=–1.9) are in close proximity to the eclipsed Sun and both will be easily visible during totality. As the brightest planet in the sky, Venus can actually be observed in broad daylight provided that the sky is cloud free and of high transparency (i.e. - no dust or particulates). During the 1998 eclipse, Venus is located 42° west of the Sun where it will reach greatest elongation the following month. Look for the planet during the partial phases by first covering the crescent Sun with an extended hand. Venus will be shining at its greatest brilliancy (m_V=–4.5) so it will be impossible to miss during totality. Mars (m_V=+1.2) and Saturn (m_V=+0.9) are located 17° and 40° east of the Sun, respectively. They will prove more challenging to spot than the other planets, but not too difficult if the sky transparency is good. Under the right circumstances, it should be possible to view all five classical planets, the Moon and the Sun (or at least it's corona) as one's eyes sweep across the darkened sky during totality.

A number of the brightest summer/autumn stars may also be visible during totality. Twenty degrees south of the Sun, Fomalhaut (m_V=+1.16) will have an altitude of 50°, while Achernar (m_V=+0.46) lies 20° above the southeastern horizon. The summer triangle composed of Altair (m_V=+0.77), Deneb (m_V=+1.25), and Vega (m_V=+0.03), will be located in the northwest while Capella (m_V=+0.08) stands 10° high in the northeast.

The following ephemeris [using Bretagnon and Simon, 1986] gives the positions of the naked eye planets during the eclipse. **Delta** is the distance of the planet from Earth (A.U.'s), **V** is the apparent visual magnitude of the planet, and **Elong** gives the solar elongation or angle between the Sun and planet.

```
Ephemeris: 1998 Feb  26    18:00:00 UT              Equinox = Mean Date

Planet      RA          Dec          Delta      V      Size    Phase   Elong

Sun         22h38m24s   -08°35'35"   0.99024   -26.7   1938.2    -       -
Mercury     22h55m01s   -08°37'19"   1.34548   -1.5       5.0   0.99    4.1E
Venus       19h47m35s   -15°56'33"   0.45918   -4.5      36.3   0.31   42.3W
Mars        23h44m19s   -02°30'47"   2.31165    1.2       4.0   0.99   17.5E
Jupiter     22h29m58s   -10°22'28"   5.99563   -1.9      32.8   1.00    2.7W
Saturn      01h09m22s   +04°52'23"  10.08826    0.9      16.4   1.00   40.0E
```

CONTACT TIMINGS FROM THE PATH LIMITS

Precise timings of beading phenomena made near the northern and southern limits of the umbral path (i.e. - the graze zones), are of value in determining the diameter of the Sun relative to the Moon at the time of the eclipse. Such measurements are essential to an ongoing project to monitor changes in the solar diameter. Due to the conspicuous nature of the eclipse phenomena and their strong dependence on geographical location, scientifically useful observations can be made with relatively modest equipment. A small telescope, short wave radio and portable camcorder are usually used to make such measurements. Time signals are broadcast via short wave stations WWV and CHU, and are recorded simultaneously as the eclipse is videotaped. If a video camera is not available, a tape recorder can be used to record time signals with verbal timings of each event. Inexperienced observers are cautioned to use great care in making such observations. The safest timing technique consists of observing a projection of the Sun rather than directly imaging the solar disk itself. The observer's geodetic coordinates are required and can be measured from USGS or other large scale maps. If a map is unavailable, then a detailed description of the observing site should be included which provides information such as distance and directions of the nearest towns/settlements, nearby landmarks, identifiable buildings and road intersections. The method of contact timing should be described in detail, along with an estimate of the error. The precisional requirements of these observations are ±0.5 seconds in time, 1" (~30 meters) in latitude and longitude, and ±20 meters (~60 feet) in elevation. Although GPS's (Global Positioning Satellite receivers) are commercially available (~$300 US), their positional accuracy of ±100 meters is about three times larger than the minimum accuracy required by grazing eclipse measurements. The International Occultation Timing Association (IOTA) coordinates observers world-wide during each eclipse. For more information, contact:

>Dr. David W. Dunham, IOTA
>7006 Megan Lane
>Greenbelt, MD 20770-3012, USA
>Phone: (301) 474-4722   E-mail: David_Dunham@jhuapl.edu

Send reports containing graze observations, eclipse contact and Baily's bead timings, including those made anywhere near or in the path of totality or annularity to:

>Dr. Alan D. Fiala
>Orbital Mechanics Dept.
>U. S. Naval Observatory
>3450 Massachusetts Ave., NW
>Washington, DC 20392-5420, USA

PLOTTING THE PATH ON MAPS

If high resolution maps of the umbral path are needed, the coordinates listed in Tables 7 and 8 are conveniently provided in longitude increments of 1° and 7.5' respectively to assist plotting by hand. The path coordinates in Table 3 define a line of maximum eclipse at five minute increments in Universal Time. If observations are to be made near the limits, then the grazing eclipse zones tabulated in Table 8 should be used. Global Navigation Charts (1:5,000,000), Operational Navigation Charts (scale 1:1,000,000) and Tactical Pilotage Charts (1:500,000) of many parts of the world are published by the Defense Mapping Agency. In October 1992, the DMA discontinued selling maps directly to the general public. This service has been transferred to the National Ocean Service (NOS). For specific information about map availability, purchase prices, and ordering instructions, contact the NOS at:

>National Ocean Service
>Distribution Branch
>N/GC33
>6501 Lafayette Avenue
>Riverdale, MD 20737, USA   Phone: 1-301-436-6990

It is also advisable to check the telephone directory for any map specialty stores in your city or metropolitan area. They often have large inventories of many maps available for immediate delivery.

ECLIPSE DATA ON INTERNET

NASA ECLIPSE BULLETINS ON INTERNET

Response to the first two NASA solar eclipse bulletins RP1301 (Annular Solar Eclipse of 1994 May 10) and RP1318 (Total Solar Eclipse of 1994 November 3) was overwhelming. Unfortunately, the demand quickly exceeded the limited number of bulletins printed with current levels of funding. To address this problem as well as allowing greater access to them, the eclipse bulletins were first made available via the Internet in April 1994. This was made possible through the efforts and expertise of Dr. Joe Gurman (GSFC/Solar Physics Branch). All future eclipse bulletins will be available via Internet.

NASA eclipse bulletins can be read or downloaded via the World-Wide Web using a Web browser (e.g.: Mosaic, Netscape, Microsoft Explorer, etc.) from the GSFC SDAC (Solar Data Analysis Center) home page:

> http://umbra.nascom.nasa.gov/sdac.html

The top-level URL for the eclipse bulletins themselves are:

http://umbra.nascom.nasa.gov/eclipse/941103/rp.html	(1994 Nov 3)
http://umbra.nascom.nasa.gov/eclipse/951024/rp.html	(1995 Oct 24)
http://umbra.nascom.nasa.gov/eclipse/970309/rp.html	(1997 Mar 9)
http://umbra.nascom.nasa.gov/eclipse/980226/rp.html	(1998 Feb 26)

The original Microsoft Word text files and PICT figures (Macintosh format) are also available via anonymous ftp. They are stored as BinHex-encoded, StuffIt-compressed Mac folders with .hqx suffixes. For PC's, the text is available in a zip-compressed format in files with the .zip suffix. There are three sub directories for figures (GIF format), maps (JPEG format), and tables (html tables, easily readable as plain text). For example, NASA RP 1344 (Total Solar Eclipse of 1995 October 24 [=951024]) has a directory for these files is as follows:

file://umbra.nascom.nasa.gov/pub/eclipse/951024/RP1344text.hqx	
file://umbra.nascom.nasa.gov/pub/eclipse/951024/RP1344PICTs.hqx	
file://umbra.nascom.nasa.gov/pub/eclipse/951024/ec951024.zip	
file://umbra.nascom.nasa.gov/pub/eclipse/951024/figures	(directory with GIF's)
file://umbra.nascom.nasa.gov/pub/eclipse/951024/maps	(directory with JPEG's)
file://umbra.nascom.nasa.gov/pub/eclipse/951024/tables	(directory with html's)

Other eclipse bulletins have a similar directory format.

Current plans call for making all future NASA eclipse bulletins available over the Internet, at or before publication of each. The primary goal is to make the bulletins available to as large an audience as possible. Thus, some figures or maps may not be at their optimum resolution or format. Comments and suggestions are actively solicited to fix problems and improve on compatibility and formats.

FUTURE ECLIPSE PATHS ON INTERNET

Presently, the NASA eclipse bulletins are published 18 to 24 months before each eclipse. This will soon be increased to 24 to 36 months or more. However, there have been a growing number of requests for eclipse path data with an even greater lead time. To accommodate the demand, predictions have been generated for all central solar eclipses from 1995 through 2005 using the JPL DE/LE 200 ephemerides. All predictions use the Moon's the center of mass; no corrections have been made to adjust for center of figure. The value used for the Moon's mean radius is $k=0.272281$. The umbral path characteristics have been predicted at 2 minute intervals of time compared to the 6 minute interval used in *Fifty Year Canon of Solar Eclipses: 1986-2035* [Espenak, 1987]. This should provide enough detail for making preliminary plots of the path on larger scale maps. Note that positive latitudes are north and positive longitudes are west.

The paths for the following fifteen eclipses are currently available via the Internet:

1995 April 29	–	Annular Solar Eclipse
1995 October 24	–	Total Solar Eclipse
1997 March 09	–	Total Solar Eclipse
1998 February 26	–	Total Solar Eclipse
1998 August 22	–	Annular Solar Eclipse
1999 February 16	–	Annular Solar Eclipse
1999 August 11	–	Total Solar Eclipse
2001 June 21	–	Total Solar Eclipse
2001 December 14	–	Annular Solar Eclipse
2002 June 10	–	Annular Solar Eclipse
2002 December 04	–	Total Solar Eclipse
2003 May 31	–	Annular Solar Eclipse
2003 November 23	–	Total Solar Eclipse
2005 April 08	–	Annular/Total Solar Eclipse
2005 October 03	–	Annular Solar Eclipse

The tables can be accessed with a Web browser through the SDAC home page, or directly at URL:

http://umbra.nascom.nasa.gov/eclipse/predictions/year-month-day.html

For example, the eclipse path of 1999 August 11 would use the above address with the string "year-month-day" replaced by "1999-august-11". Send comments, corrections, suggestions or requests for more detailed 'ftp' instructions, to Fred Espenak via e-mail (espenak@lepvax.gsfc.nasa.gov). For Internet related problems, please contact Joe Gurman (gurman@uvsp.nascom.nasa.gov).

DOWNLOADING BULLETINS AND PATH TABLES VIA ANONYMOUS FTP

The eclipse bulletins and path tables are also available via anonymous ftp for sites which do not have access to the World Wide Web. A user first ftp's to umbra.nascom.nasa.gov (150.144.30.134), using the username "anonymous" and password "<username>@<host>". Note that the password is your e-mail address where <username> is your name and <host> is the fully qualified Internet address of your machine (e.g.- gurman@uvsp.nascom.nasa.gov). Next, you change directory with the command "cd pub/eclipse".

There are four directories 941103, 951024, 970309, and 980226; one for each of the last four eclipse bulletins (1318, 1344, 1369 and 1383, respectively). In each, there is a flat ASCII README file and two .hqx files: RPnnnntext.hqx and RPnnnnPICTS.hqx, where "nnnn" is the Reference Publication number. All .hqx files are BinHex-encoded (ASCII), StuffIt-compressed files for the Macintosh. There's also one .zip file: ecyymmdd.zip, where "yymmdd" is the date of the eclipse. This is a zip-compressed and encoded file for PC's. There are also three subdirectories, figures, maps, and tables, with (respectively), the GIF figures, the JPEG GNC charts, and the html tables (easily readable as plain text). For example, the total solar eclipse of 970309 (= 1997 Mar 9) and published as NASA RP 1369 has a directory for these files is as follows:

file://umbra.nascom.nasa.gov/pub/eclipse/970309/README
file://umbra.nascom.nasa.gov/pub/eclipse/970309/RP1369text.hqx
file://umbra.nascom.nasa.gov/pub/eclipse/970309/RP1369PICTs.hqx
file://umbra.nascom.nasa.gov/pub/eclipse/970309/ec970309.zip
file://umbra.nascom.nasa.gov/pub/eclipse/970309/figures   (directory with GIFs)
file://umbra.nascom.nasa.gov/pub/eclipse/970309/maps   (directory with JPEGs)
file://umbra.nascom.nasa.gov/pub/eclipse/970309/tables   (directory with htmls)

Directories for analogous files for other solar eclipses are arranged similarly.

The html files should be downloaded in ASCII mode and the other files in binary (IMAGE) mode. If you are not using a Web viewer to access the ftp documents, you must first type either "ascii" or "binary" to download an ASCII or a binary file, respectively. You then download the file using the ftp protocol for your particular machine.

PREDICTIONS FOR ECLIPSE EXPERIMENTS

This publication has attempted to provide comprehensive information on the 1998 total solar eclipse to both the professional and amateur/lay communities. However, certain investigations and eclipse experiments may require additional information which lies beyond the scope of this work. We invite the international professional community to contact us for assistance with any aspect of eclipse prediction including predictions for locations not included in this publication, or for more detailed predictions for a specific location (e.g.: lunar limb profile and limb corrected contact times for an observing site).

This service is offered for the 1998 eclipse as well as for previous eclipses in which analysis is still in progress. To discuss your needs and requirements, please contact Fred Espenak (espenak@lepvax.gsfc.nasa.gov).

IAU WORKING GROUP ON ECLIPSES

Professional scientists are asked to send descriptions of their eclipse plans to the Working Group on Eclipses of the International Astronomical Union, so that they can keep a list of observations planned. Send such descriptions, even in preliminary form, to:

> International Astronomical Union/Working Group on Eclipses
> Jay M. Pasachoff, Chair
> Williams College-Hopkins Observatory
> Williamstown, MA 01267, USA
> fax: (413) 597-3200;   e-mail: jay.m.pasachoff@williams.edu

The members of the Working Group on Eclipses of Commissions 10 and 12 of the International Astronomical Union are: Jay M. Pasachoff (USA), Chair; F. Espenak (USA); Iraida Kim (Russia); Yoshinori Suematsu (Japan); Jagdev Singh (India); V. Rusin (Slovakia); consultant: J. Anderson (Canada).

NATIONAL ECLIPSE COMMITTEE OF VENEZUELA

Dr. Francisco Fuenmayor is the National Eclipse Committee Coordinator of Venezuela. He informs us that the University del Zulia will set up a special observing camp in Maracaibo for professionals interested in installing scientific equipment. He emphasizes that the facilities of the campground will only be available to those involved in a scientific experiment and supported by a recognized institution. If you plan to observe the eclipse from some other part in Venezuela, please contact Dr. Fuenmayor. He can give advice on any other observing site you may have selected. Scientists interested in using the special campground should fill out the following questionnaire and return it to Dr. Fuenmayor at franfuen@cida.ve as soon as possible.

-NAME OF OBSERVERS:
-NAME OF INSTITUTION:
-BRIEF DESCRIPTION OF SCIENTIFIC EXPERIMENT:
-E-MAIL ADDRESS:
-EXPECTED NUMBER OF MEMBERS IN PARTY:
-EXPECTED DATE OF ARRIVAL:
-REQUIRED AREA FOR YOUR EXPERIMENT:
-REQUIRED POWER SUPPLY:
-OTHER REQUIREMENTS:

Additional information about observing the eclipse from Venezuela has been prepared by the National Eclipse Committee. It can be obtained via ftp from: ftp chivac.cida.ve, with anonymous login. The password is your e-mail address. The directory is /pub/eclipse98, and the file is eclipse98. Those using Netscape can access the same information from the URL: ftp://chivac.cida.ve/pub/eclipse98.

The National Eclipse Committee of Venezuela includes: Dr. Francisco Fuenmayor (Coordinator) (Universidad de los Andes, ULA), Dr. Ignacio Ferrin (ULA), Javier Gonzalez (ULA), Lic. Nestor Sanchez,

Lic. Ingrid Inciarte, Prof. Jeannette Stock, Prof. Ruben Cadenas, Prof. Pedro Franceschini (Universidad del Zulia), Dra. Gladis Magris and Alberto Dubuc (Centro de Investigaciones de Astronomia, CIDA).

TOTAL SOLAR ECLIPSE OF 1999 AUGUST 11

The next total eclipse of the Sun is the final one of the twentieth century. The path of the Moon's umbral shadow begins in the North Atlantic, continues through central Europe (Figure 18), the Middle East, and south Asia, where it ends at sunset in the Bay of Bengal. A partial eclipse will be seen within the much broader path of the Moon's penumbral shadow, which includes the northeastern United States and Canada, Greenland, Iceland, all of Europe, most of Asia and the northern third of Africa [Espenak, 1987].

Europeans have been waiting for this event for 38 years. Not since 1961 has the Moon's shadow touched down on the Continent. Beginning in the Atlantic about 300 km south of Nova Scotia, the umbra quickly traverses the ocean. Southern England enjoys first landfall in Cornwall and parts of Devon. The center line duration of this mid morning eclipse is 2 minutes as the Sun stands 45 degrees above the horizon. Unfortunately, the probability of clear skies is only about 1 out of 3.

The weather prospects do not change appreciably as the path crosses the English Channel and swings through northern France. As the umbra passes 20 km north of Paris, the City of Lights will be darkened be by an eclipse of magnitude 0.994. Southern Belgium and Luxembourg also lie in the path which continues into Germany. Stuttgart is just north of center line and enjoys a total eclipse duration of 2 minutes 17 seconds. Nearly 2 million citizens of Munich will also bear witness to over two minutes of totality, provided the sky is clear on eclipse day. Traveling through central Austria and Hungary, the shadow narrowly misses Vienna and Budapest. But Bucharest, Romania, stands squarely on the center line just as the total eclipse reaches its greatest duration of 2 minutes 23 seconds.

After paralleling the Romanian/Bulgarian border, the track crosses the Black Sea and diagonally bisects Turkey. Ankara lies 150 km south of the path and experiences a partial eclipse of magnitude 0.967. The path narrows and the duration drops as shadow's trajectory takes it through Iran, Afghanistan, southern Pakistan and central India where it ends in the Bay of Bengal. Although a detailed weather study for the 1999 eclipse is not yet available, a preliminary analysis shows a clearing trend through Eastern Europe which continues to improve in Turkey, and reaching a peak in Iran.

Complete details will be published in the next NASA solar eclipse bulletin in Fall-Winter 1996.

ALGORITHMS, EPHEMERIDES AND PARAMETERS

Algorithms for the eclipse predictions were developed by Espenak primarily from the *Explanatory Supplement* [1974] with additional algorithms from Meeus, Grosjean and Vanderleen [1966] and Meeus [1982]. The solar and lunar ephemerides were generated from the JPL DE200 and LE200, respectively. All eclipse calculations were made using a value for the Moon's radius of $k=0.2722810$ for umbral contacts, and $k=0.2725076$ (adopted IAU value) for penumbral contacts. Center of mass coordinates were used except where noted. Extrapolating from 1996 to 1998, a value for ΔT of 63.4 seconds was used to convert the predictions from Terrestrial Dynamical Time to Universal Time. The international convention of presenting date and time in descending order has been used throughout the bulletin (i.e. - *year, month, day, hour, minute, second*).

The primary source for geographic coordinates used in the local circumstances tables is *The New International Atlas* (Rand McNally, 1991). Elevations for major cities were taken from *Climates of the World* (U. S. Dept. of Commerce, 1972).

All eclipse predictions presented in this publication were generated on a Macintosh Quadra 630 computer. Word processing and page layout for the publication were done using Microsoft Word v5.1. Figures were annotated with Claris MacDraw Pro 1.5. Meteorological diagrams were prepared using Corel Draw 3.0 and converted to Macintosh compatible files. Finally, the bulletin was printed on a 600 dpi laser printer (Apple LaserWriter Pro).

The names and spellings of countries, cities and other geopolitical regions are not authoritative, nor do they imply any official recognition in status. Corrections to names, geographic coordinates and elevations are actively solicited in order to update the data base for future eclipses. All calculations, diagrams and opinions presented in this publication are those of the authors and they assume full responsibility for their accuracy.

BIBLIOGRAPHY

REFERENCES

Bretagnon, P., and Simon, J. L., *Planetary Programs and Tables from −4000 to +2800*, Willmann-Bell, Richmond, Virginia, 1986.
Chou, B. R., "Safe Solar Filters," *Sky & Telescope*, August 1981, p. 119.
Climates of the World, U. S. Dept. of Commerce, Washington DC, 1972.
Dunham, J. B, Dunham, D. W. and Warren, W. H., *IOTA Observer's Manual,* (draft copy), 1992.
Espenak, F., *Fifty Year Canon of Solar Eclipses: 1986–2035*, NASA RP-1178, Greenbelt, MD, 1987.
Explanatory Supplement to the Astronomical Ephemeris and the American Ephemeris and Nautical Almanac, Her Majesty's Nautical Almanac Office, London, 1974.
Herald, D., "Correcting Predictions of Solar Eclipse Contact Times for the Effects of Lunar Limb Irregularities," *J. Brit. Ast. Assoc.*, 1983, **93**, 6.
Marsh, J. C. D., "Observing the Sun in Safety," *J. Brit. Ast. Assoc.*, 1982, **92**, 6.
Meeus, J., *Astronomical Formulae for Calculators,* Willmann-Bell, Inc., Richmond, 1982.
Meeus, J., Grosjean, C., and Vanderleen, W., *Canon of Solar Eclipses*, Pergamon Press, New York, 1966.
Morrison, L. V., "Analysis of lunar occultations in the years 1943–1974...," *Astr. J.*, 1979, **75**, 744.
Morrison, L.V., and Appleby, G.M., "Analysis of lunar occultations - III. Systematic corrections to Watts' limb-profiles for the Moon," *Mon. Not. R. Astron. Soc.*, 1981, **196**, 1013.
The New International Atlas, Rand McNally, Chicago/New York/San Francisco, 1991.
van den Bergh, G., *Periodicity and Variation of Solar (and Lunar) Eclipses*, Tjeenk Willink, Haarlem, Netherlands, 1955.
Watts, C. B., "The Marginal Zone of the Moon," *Astron. Papers Amer. Ephem.*, 1963, **17**, 1-951.

METEOROLOGY AND TRAVEL

Cameron, S. and Box, B., *Caribbean Islands Handbook*, Passport Books, Lincolnwood (Chicago), 1995.
Constance, P., *The Galapagos Islands*, Passport Books, Lincolnwood (Chicago), 1995.
Garcia, O., *Atlas of Highly Reflective Clouds for the Global Tropics*, U.S. Department of Commerce, Boulder, 1985.
U.S. Naval Weather Service Command, *Summary of Synoptic Meteorological Observations: Caribbean and Nearby Island Coastal Marine Areas* (microfiche), National Technical Information Service, Springfield, VA, 1978.

FURTHER READING

Allen, D., and Allen, C., *Eclipse*, Allen & Unwin, Sydney, 1987.
Astrophotography Basics, Kodak Customer Service Pamphlet P150, Eastman Kodak, Rochester, 1988.
Brewer, B., *Eclipse*, Earth View, Seattle, 1991.
Covington, M., *Astrophotography for the Amateur*, Cambridge University Press, Cambridge, 1988.
Espenak, F., "Total Eclipse of the Sun," *Petersen's PhotoGraphic*, June 1991, p. 32.
Fiala, A. D., DeYoung, J. A., and Lukac, M. R., *Solar Eclipses, 1991–2000*, USNO Circular No. 170, U. S. Naval Observatory, Washington, DC, 1986.
Harris, J., and Talcott, R., *Chasing the Shadow,* Kalmbach Pub., Waukesha, 1994.
Littmann, M., and Willcox, K., *Totality, Eclipses of the Sun*, University of Hawaii Press, Honolulu, 1991.
Lowenthal, J., *The Hidden Sun: Solar Eclipses and Astrophotography*, Avon, New York, 1984.
Mucke, H., and Meeus, J., *Canon of Solar Eclipses: −2003 to +2526*, Astronomisches Büro, Vienna, 1983.
North, G., *Advanced Amateur Astronomy*, Edinburgh University Press, 1991.
Oppolzer, T. R. von, *Canon of Eclipses*, Dover Publications, New York, 1962.
Ottewell, G., *The Under-Standing of Eclipses*, Astronomical Workshop, Greenville, NC, 1991.
Pasachoff, J. M., and Covington, M., *Cambridge Guide to Eclipse Photography*, Cambridge University Press, Cambridge and New York, 1993.
Pasachoff, J. M., and Menzel, D. H., *Field Guide to the Stars and Planets*, 3rd edition, Houghton Mifflin, Boston, 1992.
Reynolds, M. D. and Sweetsir, R. A., *Observe Eclipses*, Astronomical League, Washington, DC, 1995.
Sherrod, P. C., *A Complete Manual of Amateur Astronomy*, Prentice-Hall, 1981.

Zirker, J. B., *Total Eclipses of the Sun*, Princeton University Press, Princeton, 1995.